SOUTHERN BY-WAYS: BRANCH LINES OF BR'S SOUTHERN REGION IN THE 1960s

LAURIE GOLDEN

First published 2024

Amberley Publishing
The Hill, Stroud
Gloucestershire, GL5 4EP

www.amberley-books.com

Copyright © Laurie Golden, 2024

The right of Laurie Golden to be identified as the Author of this work has been asserted in accordance with the Copyrights, Designs and Patents Act 1988.

ISBN 978 1 3981 1257 5 (print)
ISBN 978 1 3981 1258 2 (ebook)

All rights reserved. No part of this book may be reprinted or reproduced or utilised in any form or by any electronic, mechanical or other means, now known or hereafter invented, including photocopying and recording, or in any information storage or retrieval system, without the permission in writing from the Publishers.

British Library Cataloguing in Publication Data.
A catalogue record for this book is available from the British Library.

Origination by Amberley Publishing.
Printed in the UK.

Introduction

The Southern Railway was formed at 'grouping' in 1923 primarily from three constituent companies – the London & South Western Railway (LSWR), the London, Brighton & South Coast Railway (LBSCR) and the South Eastern & Chatham Railway (SECR), itself formed of the South Eastern Railway (SER) and the London, Chatham & Dover Railway (LCDR) in 1899. This combined system became the Southern Region of British Railways (BR) at nationalisation in 1948.

By 1963, when I started colour photography, most locomotives of the constituent companies had been withdrawn from service or were on their last legs, confined to secondary routes and branch lines.

By this time the by-ways in Kent had all either been closed or turned to modern traction, so we start at (Royal) Tunbridge Wells, which is just inside Kent on the border with Sussex (now East Sussex). Two lines of interest radiated from here. Diverging at Groombridge Junction, in roughly southerly and north-westerly directions, the Tunbridge Wells–Eastbourne and Tunbridge Wells–Oxted were still steam operated in 1963.

The former, built by the LBSCR starting in 1848 and finally opening to its full length in 1890, passed through rural Sussex. By 1963, the line was operated almost exclusively by BR 4MT 2-6-4Ts. The Oxted line, opened in the 1880s by the LBSCR, was served from Victoria or London Bridge. It passed through Surrey to Oxted and then meandered across the border of Sussex and Kent before arriving at Ashurst Junction where it diverged eastwards to Tunbridge Wells. In 1963, it was still operated by push-pull (motor) trains using pre-grouping tank engines.

In this type of operation, the locomotive remained at the same end of the coaching stock whichever direction the train was travelling. The stock was modified with a driving cab, providing the driver with basic controls at the cab end along with a bell or other signalling code system to communicate with the fireman (located in the engine itself) and pass commands to adjust controls not available in the cab.

At the time of these illustrations, the Oxted line was still a double-track secondary through route to Brighton, bypassing Tunbridge Wells by way of Ashurst and Birchden junctions. It was quite unusual to see typical branch line push-pull units on a double-track formation. The route still exists today, but only as far as Uckfield, and with no direct connection to Tunbridge Wells. However, the heritage Spa Valley Railway today operates a steam train service between the former Tunbridge Wells West and Eridge using the former Down track from Birchden Junction.

Moving westwards, we come to the Three Bridges to Tunbridge Wells line, which opened in 1855 and was operated by the LBSCR. Running in a west to east direction, this line was, at this time, also operated by push-pull steam trains until the end of 1963. The line closed completely in June 1965.

Some 15 miles north of Three Bridges is the Surrey town of Redhill, from which the North Downs line runs west to Guildford along the foot of the North Downs and then onwards in a north-westerly direction to Reading. It was opened in 1849 and was operated by the SER.

A number of short sections of the route had been electrified: the first couple of miles from Redhill to Reigate, the section from Guildford to Ash, and the section to Reading from Wokingham, where the line from Waterloo via Ascot joined the route. However, train services between Redhill and Reading were steam hauled until the end of 1964 with a variety of motive power. The line is still open and forms an important link to Gatwick Airport, which is located just north of Three Bridges.

The North Downs line joined the main London (Waterloo) to Portsmouth line 2 miles south of Guildford at Shalford Junction. Less than a mile further south, at Peasmarsh Junction, the Guildford to Horsham line headed in a south-easterly direction to join the Three Bridges to Arundel line at Christs Hospital, 3 miles south of Horsham. The line was built and operated by the LBSCR in 1865. It was steam operated right until its closure in June 1965, a few months short of its centenary.

The 'new line' from Hampton Court Junction to Guildford via Effingham Junction was opened by the LSWR in 1885, and was fully electrified in 1925. Fortunately, an 'as required' steam-hauled freight train ran from Surbiton to Cobham every Tuesday lunchtime, enabling me to dash over from my place of work near Chiswick during my lunch break.

The Alton line through Aldershot branched off at Pirbright Junction, just west of Brookwood on the Bournemouth main line. At Alton, this made an end-on connection to the Mid-Hants (Watercress) line which rejoined the Bournemouth line near Winchester, making it at the time a useful diversionary route. It was fully opened in 1870, and the section up to Alton was electrified in 1937.

The section from Alton to Winchester was never electrified, and by the time covered by this book passenger services were in the hands of Hampshire diesel multiple units (DMUs). As a result of Sunday closures of the Bournemouth line on a number of occasions in 1965/66 due to work on the conversion to electrification, it was still possible to photograph steam trains on the line. Due to the 4-mile 1 in 60 climb from Alton to Medstead, the heavier westbound trains were double-headed. The line from Alton to Winchester closed in 1973, but the section from Alton to Arlesford is now operated by the heritage Mid-Hants Railway.

The route from Salisbury to Romsey and Eastleigh was opened by the LSWR in 1847, and the dead straight line from Eastleigh to Fareham was opened by the LSWR in 1841. At Fareham, it joins the line running along the south coast from Brighton to Southampton, the section from Fareham to Southampton being fully opened by the LSWR in 1889. Apart from through trains from Brighton to Bristol and Cardiff, by 1963 passenger services were mainly in the hands of modern traction, but Sunday diversions from Eastleigh to Southampton via Fareham in 1964 due to the engineering work on the main line between Eastleigh and Southampton saw main-line steam activity.

Havant is a town some 10 miles east of Fareham, and from here a branch line ran to Hayling Island. This is a true island, surrounded on all sides by water, and linked to the mainland by a road causeway and until November 1963 a railway viaduct. The line was constructed by the LBSCR and opened for passengers in 1867. It ran down the west side of the island to West Town. Due to weight restrictions on this timber viaduct, the branch was worked by LBSCR A1/A1X 0-6-0s for almost its entire life.

The Hampshire New Forest town of Brockenhurst is the junction for the 6-mile-long branch line to Lymington, which was opened by the LSWR in 1858. Due to the important ferry link from Lymington to Yarmouth on the Isle of Wight, the branch is still open, but now electrified as part of the Bournemouth electrification scheme. The last steam train ran on 2 April 1967.

Between Bournemouth and Bath, the Somerset & Dorset Joint Railway (SDJR) was formed in 1875 jointly by the Midland Railway and the LSWR, after an amalgamation of the Somerset

Central and Dorset Central railways. This performed a useful through route for traffic between the Midlands and the north of England, and the south coast, and on summer Saturdays it handled a considerable volume of holiday trains. After nationalisation, control was entirely in the hands of the Southern Region of British Railways, but when the section north of Templecombe fell into the hands of the Western Region in 1958, services were allowed to decline until the line finally closed in March 1966. Blandford station remained open until 1969 for freight only as a branch from Broadstone.

The Swanage branch from Wareham opened in 1885, operated by the LSWR. Until 1966 the branch was steam operated, but even with the replacement diesel service the line was eventually closed at the end of 1971. Luckily, that was not the end of the story because the Swanage Railway Society was formed and by 1995 services were restored between Swanage and Norden Park & Ride, just north of Corfe Castle. It has been subsequently linked back to the main line at Worgret Junction with the hope of restoring services back to Wareham. From 2009, occasional excursion trains have run the full length of the branch from the main line.

A note about the M7s illustrated in this book. The M7 Class 0-4-4T was introduced by Drummond for the LSWR in 1897. Although normally still classified as M7, Drummond introduced a development in 1903 with longer overhang at the front end, which were classified at the time as X14. This longer overhang provided space to fit an air compressor to operate the push-pull system, so it was only the X14 Class that were fitted for push-pull operation. Therefore, all the M7s seen in this book were actually X14s.

My interest in trains started when I attended Surbiton Grammar School, which was located very close to the Southern main line. I started train photography in 1955 when I was about sixteen, processing black-and-white film myself, and moved to colour in 1963. I continue with still and video photography of British steam trains to this day. All images in this book, except the last, were taken with a Periflex 3 camera using Perutz Colour film.

A number of images in this book were taken from the trackside. To facilitate this I had a lineside pass, which gave me access to all non-electrified lines of the Southern Region.

Tunbridge Wells to Eastbourne and Oxted

BR 4MT 2-6-4T No. 80019 emerges from Grove Hill Tunnel, Tunbridge Wells, with the 08.38 from Tonbridge to Tunbridge Wells West. With flat-sided stock, the train could pass through the narrow-bore Groom Tunnel between here and the West station.

A BR 4MT 2-6-4T leaves the 1867 single-track connection between Tunbridge Wells West and the main Tonbridge to Hastings line at Grove Junction, heading towards Tunbridge Wells Central and Tonbridge.

Ex-SECR H Class 0-4-4T No. 31518 waits to depart Tunbridge Wells West with the 15.00 for Oxted. The locomotive shed is in the background. This location is today the headquarters of the Spa Valley Railway, which runs heritage trains from here through Groombridge to Eridge.

Ex-SECR H Class 0-4-4T No. 31518 leaves Groombridge station with the 10.00 Tunbridge Wells West to Oxted auto-train in April 1963. Just beyond here, this train will branch northwards at Groombridge Junction, whereas the train for Eastbourne will take the southern spur to Birchden Junction.

An unidentified BR Class 4MT tank leaves the bay platform at Eridge with the 19.14 short working to Tunbridge Wells West. Steam trains can still be seen here on a regular basis since Eridge is the terminus of the Spa Valley Railway.

BR Class 4MT 2-6-4T No. 80034 pauses at Mayfield, the second station down the line from Redgate Mill Junction, where the Eastbourne line diverged from the Brighton line (still open to Uckfield). With wide-bodied set 863, this train would terminate at Tunbridge Wells West.

BR Class 4MT 2-6-4T No. 80152 is passing through Markly Wood, north of Heathfield, with the 09.45 from Eastbourne to Tunbridge Wells. This loco was one of the batch of this class built at Brighton Works.

BR Class 4MT 2-6-4T No. 80033 emerges from the short Heathfield Tunnel into the station with narrow-bodied set 537, the 11.08 from Tonbridge. Named 'Cuckoo Line' by the railwaymen themselves, this relates to the old Sussex legend that annually on 14 April the first cuckoo of summer is released at Heathfield Fair.

BR Class 4MT 2-6-4T No. 80032 is passing through Horam with a short freight from Polegate to Heathfield. Horam station is located right in the centre of the village.

The number is unreadable on this filthy BR Class 4MT 2-6-4T as it pulls into Horam station with the 13.39 train from Tunbridge Wells West to Eastbourne.

The signalman stands with the single-line staff from Hailsham, which the crew of BR Class 4MT 2-6-4T No. 80141 on the 15.45 from Eastbourne to Tunbridge Wells West will have just handed over.

Hellingly station, sited between Horam and Hailsham, was the only single-platform station on the line. BR Class 4MT 2-6-4T No. 80032 returns from Heathfield with an even shorter freight. The land in the foreground was once occupied by a short siding leading to Hellingly Mental Hospital, operated by a twelve-seat four-wheeled tramcar.

BR Class 4MT 2-6-4T No. 80144 leaves Hailsham with the 17.41 from Tunbridge Wells West to Eastbourne. Hailsham, about 3 miles north of the junction of the Cuckoo Line with the main line from Brighton to Eastbourne, was the largest town on the line.

The relative importance of Hailsham was demonstrated by the fact that it justified short workings to and from Eastbourne at rush hour times. One such train, the 18.22 from Eastbourne hauled by a BR Class 4MT 2-6-4T, is approaching the station.

An unidentified BR Class 4MT 2-6-4T arrives at Polegate with the 19.06 working from Hailsham. Polegate was the junction with the main line from Brighton to Hastings.

Ex-SR U Class 2-6-0 No. 31803 and N Class 2-6-0 No. 31411 are passing Hailsham with the LCGB Wealdsman rail tour on Sunday 13 June 1965, the last day of operation along the full length of the Cuckoo Line, although a diesel service between Polegate and Hailsham continued for another three years. This trip marked the closure of a number of branch lines in the area, which we will see further.

Ex-LSWR M7 Class 0-4-4T No. 30055 heads northwards from Groombridge Junction to join the Oxted–Brighton line at Ashurst Junction en route to Oxted with the 18.00 from Tunbridge Wells West.

Bathed in late afternoon sunshine earlier the same day, ex-LSWR M7 Class 0-4-4T No. 30055 pauses at Ashurst station with the 16.04 from Oxted to Tunsbridge Wells West. By 1963 the fleet of ex-SECR H Class locos was diminishing, so one or two M7s had been drafted in to assist.

Ex-SECR H Class 0-4-4T No. 31544 is arriving at Cowden with the 11.04 from Oxted, bound for Tunbridge Wells West. The village of Cowden is over a mile away from the station with its name.

Ex-SECR H Class 0-4-4T No. 31005 emerges from Markbeech Tunnel to arrive at Cowden with the 13.04 from Oxted to Tunbridge Wells West.

Ex-SECR H Class 0-4-4T No. 31544 leaves Hever bound for Tunbridge Wells West with the 15.04 from Oxted. It was uncommon to find the locomotive coupled to the stock chimney first when working push-pull.

Ex-SECR H Class 0-4-4T No. 31005 is just departing Hever station with the 16.40 from Oxted to Tunbridge Wells. About a mile to the East lies Hever Castle, once owned in the sixteenth century by the Boleyn family and home of Anne Boleyn, the infamous second wife of Henry VIII.

Ex-SECR H Class 0-4-4T No. 31518 is pictured leaving Edenbridge Tunnel bound for Oxted with the 16.00 from Tunbridge Wells West. The line from Redhill to Tonbridge crosses above the tunnel. Both lines are still operating today.

Ex-SECR H Class 0-4-4T No. 31551 is entering Oxted station with the 18.00 from Tunbridge Wells West. The trains from Tunbridge Wells West via Hever terminated at Oxted, but during most of the day connected with a through train to London Victoria that had left Tunbridge Wells West some 15 minutes earlier, looping round via Forest Row and East Grinstead to arrive at Oxted 10 minutes after the Hever line train.

Three Bridges to East Grinstead

Ex-SECR H Class 0-4-4T No. 31263 is waiting in the bay platform at Three Bridges to depart at 11.08 for East Grinstead. Three Bridges was the junction with the London to Brighton main line and is also still a junction for trains to Horsham and Littlehampton.

Ex-SECR H Class 0-4-4T No. 31263 is working hard up the 1 in 88 bank between Three Bridges and Rowfant with the 12.08 from Three Bridges. Push-pull trains on this line outlived those on the Tunbridge Wells–Oxted service by around six months.

Ex-SECR H Class 0-4-4T No. 31543 is arriving at Rowfant bound for East Grinstead. Shell and BP constructed semi-buried air force reserve fuel tanks at Rowfant station during the Second World War. They were transferred to the Ministry of Power for storage of jet fuel in 1959.

Ex-SECR H Class 0-4-4T No. 31518 enters Rowfant station on the 11.08 from Three Bridges in the summer of 1963. By this time, the wartime fuel tanks, although extant, appeared to be out of use.

Ex-SECR H Class 0-4-4T No. 31543 is heading west from East Grinstead with the 18.27 train for Three Bridges, illuminated by the evening sunshine.

Ex-LSWR M7 Class 0-4-4s also found their way onto this branch, as we see No. 30029 on the approach to East Grinstead with the 12.08 from Three Bridges.

On a rather dull day, ex-SECR H Class 0-4-4T No. 31543 pushes its train into East Grinstead High Level station, which it shared with trains from London. There is also a Low Level station at right angles which was (and now again is) for the Bluebell Line.

The last day of steam service was marked with the LCGB Wealdsman rail tour on Sunday 13 June 1965 using ex-SR U Class 2-6-0 No. 31803 and N Class 2-6-0 No. 31411, seen leaving Rowfant station. Since it was the custom to clean locos used on special trains such as this, the poor external condition of No. 31803 suggests this was a late substitution for a failed loco. The out of use siding to the former fuel dump can be seen in the foreground.

Redhill to Reading

Viewed across a line of condemned wagons adjacent to the locomotive shed, a BR Class 4MT 2-6-4T hauls a train from Reading round the curve into Redhill station. Any trains continuing towards Brighton or Tonbridge had to reverse in the station.

Mainly in the summer months, interregional trains arrived at Redhill with 'foreign' motive power, normally LMS Black 5s. In late summer 1964, considerable interest was sparked by the arrival of an ex-LNER B1 4-6-0 No. 61313, particularly as it had failed at Redhill with a hot box and was stuck on shed for several weeks. It is here working back north after repair on the Quarry Line, which bypasses Redhill station, with a special freight on 3 October.

An unidentifiable ex-GWR Manor Class 4-6-0 is approaching Betchworth working the 10.18 Redhill to Reading in January 1964. A morning train was regularly worked by an engine from Reading Western Region shed for crew familiarisation purposes.

At Dorking Town station is ex-SR U Class 2-6-0 No. 31790, doyen of the class. This was one of the twenty locos rebuilt in 1928 from the ill-fated K or River Class 2-6-4Ts, in this case A790 *River Avon*, after a number of minor derailments and finally the fatal Sevenoaks derailment which cost thirteen lives.

Ex-SECR N Class 2-6-0 No. 31811 is approaching Gomshall & Shere with the 11.35 from Redhill. The main difference between the N and U classes was driving wheel diameter (5 feet 6 inches vs. 6 feet), the N being designed as more suitable for freight work.

Ex-SR U Class 2-6-0 No. 31803, rebuilt from River Class 2-6-4T A803 *River Itchen*, is at the summit of Albury Bank between Dorking and Gomshall & Shere in the summer of 1964 with the 16.04 from Redhill.

Ex-SECR N Class 2-6-0 No. 31862 is arriving after a frosty night at Shalford station on the Guildford to Redhill North Downs line with the 09.45 Reading–Redhill in January 1964.

The same train, with ex-SR N Class 2-6-0 No. 31862, makes a spirited departure from Shalford having picked up the passengers seen standing at the station. Although primarily designed for freight work, the N Class was equally at home on passenger duties.

An ex-SR N Class 2-6-0 is on the Portsmouth direct line from Guildford approaching Shalford Junction, where it will diverge eastwards on the North Downs line to Redhill with the 14.50 from Reading in January 1964.

Ex-SR N Class 2-6-0 No. 31858 emerges from Guildford Tunnel into the winter sunlight heading for Shalford Junction and the line to Redhill with the 13.50 from Reading. This was one of a batch of fifty locos built for the SR by the Royal Arsenal at Woolwich in 1924.

A BR 4MT Class 4-6-0 departs a snowy Guildford station with the 10.55 to Redhill in January 1964. Guildford was (and still is) an important railway hub, with electric trains from Waterloo on the 'new line' through Claygate and on the Portsmouth direct line from Woking, as well as the Reading–Redhill line.

As the BR 4MT 4-6-0 pictured in the previous image accelerates away from the platform at Guildford, a BR Class 4MT 2-6-4T arrives with the 09.45 from Reading. The 'third rail' electrification is very obvious in this picture.

At a warmer time of the year, ex-SR U Class 2-6-0 No. 31791, rebuilt from River Class 2-6-4T A791 *River Adur*, leaves Guildford at 11.42 with the daily 08.45 through train from Margate to Wolverhampton.

BR 4MT Class 2-6-4T No. 80034 passes the advanced starter signal at Guildford as it curves round on the line towards Ash with the 12.37 from Guildford to Reading in the summer of 1964. As can be seen, this section of line to Ash is electrified for services from Guildford to Aldershot, Farnham and Alton.

Ex-SR N Class 2-6-0 No. 31405 is passing Ash Junction with the 12.32 from Redhill to Reading in the summer of 1964. I am standing on the clearly long-disused line to Farnham via Tongham, which was closed to passengers in 1937 but retained for freight to Tongham until the end of December 1960. The loco is one of the last batch built by the SR in 1932–34.

Ex-SECR N Class 2-6-0 No. 31407 is passing the site of Ash Junction with the 13.50 Reading–Redhill train in May 1964. The concrete plate-layer's hut occupies the site of the original junction signal box, which had been demolished the previous year.

An ex-SR N Class 2-6-0 hauls a train of petrol tankers from Farnborough North across the Basingstoke Canal in the winter of 1964. This canal was constructed to run from Basingstoke to the Wey Navigation at Byfleet. At the time of this photograph, the canal was largely disused, but has subsequently been restored for navigation.

Shortly beyond the canal crossing the electrified line to Aldershot branched off. Steam and smoke tower into the atmosphere as an-ex SR Q1 Class 0-6-0 in deplorable condition leaves Farnborough North Camp with a train of oil tankers in the winter of 1964.

Ex-SR N Class 2-6-0 No. 31831 runs into North Camp station, Farnborough, with a Christmas parcels train in December 1963. This is one of three stations in Farnborough: two on the Guildford–Reading line and one on the main line between Woking and Basingstoke.

Ex-SR U Class 2-6-0 No. 31790, rebuilt from the first River Class 2-6-4T, *River Avon*, leaves Farnborough North (the second station in the town on the Reading line) with the 15.35 from Reading to Redhill.

Ex-SR N Class 2-6-0 No. 31412 hurries along between Crowthorne and Sandhurst with the 16.20 from Reading in September 1964. Sandhurst is the location of the Royal Military Academy.

Ex-SR U Class 2-6-0 No. 31622 leaves Crowthorne station with the 17.05 from Reading in September 1964. This was a newly built loco by the Southern Railway and was not converted from a River Class tank engine.

Viewed from the signal box seen in the previous image, ex-SR N Class 2-6-0 No. 31816 starts the 15.35 from Reading away from Crowthorne.

The line between Crowthorne and Wokingham is completely straight for several miles, and ex-SR N Class 2-6-0 No. 31850 is heading south along this stretch with the 16.20 from Reading.

While the Maunsell moguls were the predominant class on the line, more modern BR classes would put in an occasional appearance, as seen here. BR 4MT No. 75074 enters Wokingham station with the 14.50 from Reading. At this point, the track is electrified after the junction with the line from Ascot.

The Ascot line diverged shortly south of Wokingham station, and the track can just be made out in front of the repeater signal on the right of the picture. BR Class 3MT 2-6-0 No. 76033 departs with the 15.35 from Reading. Wokingham church's spire is prominent in the background.

Although the passenger service on the Ascot line was entirely electric multiple units, steam could still be found on interregional freight services from Feltham yard. Ex-SR S15 Class 4-6-0 No. 30837 is passing Longcross Halt, between Staines and Ascot, with the regular morning freight.

Ex-SR S15 Class 4-6-0 No. 30837 is returning near Earley with the afternoon freight to Feltham. On these freights, the SR loco would be substituted by WR motive power at Reading. No. 30837 is pictured in much cleaner condition in a later image (see p. 51).

BR Class 4MT 2-6-4T No. 80089 departs from Reading (Southern) station with the 09.45 for Redhill. At this time the SR line from Guildford terminated in a dedicated station alongside, but at a slightly lower level than the WR main-line station – some vans can be seen on it on the right of the picture. The SR station is now closed, and trains run into the WR station.

The sunset of steam on the North Downs line! A BR 4MT Class 2-6-4T approaches Dorking Town station with the 13.50 from Reading on the snowy afternoon of 29 December 1964, just before the end of regular steam services on the line.

Guildford to Horsham

Until June 1965 another steam service was based at Guildford, namely on the line to Horsham. Ex-LMS Ivatt 2MT 2-6-2T No. 41301 leaves Guildford with the 10.34 to Horsham in autumn 1964.

About half a mile on the Portsmouth direct line beyond Shalford Junction, where the Redhill line diverged, was Peasmarsh Junction, where the Horsham line left in a south-easterly direction. Here, ex-SR Q1 0-6-0 No. 33017 takes the junction with the 18.34 to Cranleigh.

Shortly after Peasmarsh Junction, the line crossed the River Wey, and here an unidentified ex-LMS Ivatt 2MT 2-6-2T crosses with the 10.34 from Guildford to Horsham on 12 June 1965, the last day of regular passenger service.

On a bright but cold day in March, ex-LMS Ivatt 2MT 2-6-2T No. 41321 is running between Bramley and Cranleigh with the 17.04 from Guildford to Horsham. By this late stage of its life the service on the line was sparse, this being the first train from Guildford since the 10.34 a.m.

Cranleigh was the major station on the line, and here ex-LMS Ivatt 2MT 2-6-2T No. 41287 runs in with the 10.34 train from Guildford to Horsham. Cranleigh has grown considerably in population since the line was closed, and today a commuter service into Guildford would have been very useful to reduce congestion on the local roads.

The same train is seen departing Cranleigh. With closure in the offing, this view shows everything getting into a very run-down condition.

Even at this stage in the life of the branch, it was considered necessary to run a daily 'short' service to Cranleigh – the 09.08 from Guildford. In autumn 1964, ex-SR Q1 Class 0-6-0 No. 33012 'brews up' prior to departure to Baynards, where it would run round the train for the return journey.

Baynards station, built solely for the owner of Baynards Park, was remarkably large and well-appointed considering there was virtually no population within a mile. Ex-LMS Ivatt 2MT 2-6-2T No. 41301 excites a young passenger as it pauses with the 18.15 from Horsham.

As part of the route of the joint RCTS/LCGB 'Midhurst Belle' on 18 October 1964, it was hauled along the Guildford–Horsham line by ex-SR USA 0-6-0T No. 30064 seen arriving at Baynards – almost certainly the first (and last) time this class of locomotive would have worked along this line. These US Army Transportation Corps locos were purchased by the SR after the war and were used primarily in Southampton Docks and other locations for shunting.

Ex-LMS Ivatt 2MT 2-6-2T No. 41299 enters Rudgwick on the 16.53 Horsham to Guildford train in 1964. This station was located just over a mile from Baynards – further illustration of the lack of need for a station at the former point.

Christ's Hospital was the point at which the line from Guildford joined the electrified line from Three Bridges to Arundel for the final 3 miles into Horsham. Ex-LMS Ivatt 2MT 2-6-2T No. 41301 takes the branch to enter the station with the 15.09 from Horsham.

The same train departs the well-appointed station, which served the public school for boys of the same name founded by Henry VIII.

A final view of the LCGB Wealdsman tour of 13 June 1965, with ex-SR Q1 Class 0-6-0 Nos 33006 and 33027 leaving a photographic stop at Baynards. Despite its remote location, the Thurlow Arms pub was adjacent, and it was noticed that a number of train passengers availed themselves of a quick libation while others photographed the train. The Thurlow Arms, built as a railway hotel, closed in 2009, forty-four years after the railway.

The 'New Line'

The Tuesday-only goods train from Surbiton to Cobham in 1964 passes through Claygate station headed by ex-SR U Class 2-6-0 No. 31799, rebuilt from River Class 2-6-4T A799 *River Test*.

The more usual motive power for this train was an ex-SR Q1 Class 0-6-0, and here on another occasion No. 33009 passes through Claygate station. The train ran at lunchtime and was a convenient little trip when I worked at Chiswick.

Ex-SR Q1 Class 0-6-0 No. 33018 is approaching Cobham in April 1965. The front numberplate has been cleaned to portray its pre-nationalisation number, C18, which used Bulleid's unique numbering system to portray the wheel formation (C = three sets of driving wheels).

This numbering is displayed more clearly as C18 pauses during shunting at Cobham. This classification also resulted in the class of engines being known as 'Charlies' or 'Uncle Charlies'.

Pirbright Junction to Alton and Winchester

An unidentified ex-SR N Class 2-6-0 approaches Tunnel Hill, having just left the Southern main line at Purbright Junction, near Brookwood, with a Christmas parcels train for Alton in December 1964.

Ex-SECR N Class 2-6-0 No. 31816 pauses at Aldershot with a Christmas parcels train from Woking to Alton in December 1964. These Christmas parcels services were of considerable interest to enthusiasts since they often worked routes where steam was usually uncommon.

Ex-SR Q1 Class 0-6-0 No. 33027 leaves Aldershot with a Christmas parcels train for Alton in December 1965. Although not the case here, these parcels trains sometimes employed rarely used or stored locos.

Ex-SR MN Class 4-6-0 No. 35010 *Blue Star* comes off Purbright Junction, near Brookwood, with the 18.30 Waterloo–Bournemouth on 24 April 1966, being diverted via Alton. The Up line to the junction is in the foreground, curving to fly over the main line.

Ex-SR WC Class 4-6-2 No. 34002 *Salisbury* heads the 11.30 Waterloo–Weymouth at Butts Junction on 1 May 1966. The line in the foreground is the truncated Meon Valley branch. This closed to passengers in 1955, but remained open for freight between Alton and Farringdon until 1968.

Both in deplorable external condition, BR Class 3 2-6-0 No. 77014 and WC 4-6-2 No. 34102 *Lapford* climb Medstead Bank near Butts Junction with the 12.30 Waterloo–Bournemouth (*Bournemouth Belle*) on 18 September 1966. The BR Class 3 2-6-0s were normally based in the north-east of England, but No. 77014 had been transferred to Guildford and was regarded as a bit of a celebrity, so was specially diagrammed for this working, being attached at Alton.

Southern By-Ways: Branch Lines of BR's Southern Region in the 1960s   49

Ex-SR MN Class 4-6-2 No. 35008 *Orient Line* makes a spirited unassisted climb of Medstead Bank with eleven-plus coaches on the 10.30 Waterloo–Weymouth on 1 May 1966.

Again unassisted, but with a significantly lighter load, ex-SR BB Class 4-6-2 No. 34060 *25 Squadron* again storms up Medstead Bank with the 10.30 Waterloo–Weymouth on 24 April 1966.

Ex-SR BB Class 4-6-2 No. 34059 *Sir Archibald Sinclair* passes through Medstead station with the 10.34 Bournemouth–Waterloo on 15 May 1966. During the Battle of Britain, Sinclair was Secretary of State for Air.

Ex-BR Class 5MT 4-6-0 No. 73170 passes through Ropley with the 14.25 Bournemouth–Waterloo on 15 May 1966.

Ex-SR U Class 2-6-0 No. 31639 and Q1 Class 0-6-0 No. 33006 are passing though Ropley with the LCGB Wilts and Hants tour on 3 April 1966 on its return to Waterloo. Note the topiary on the platform, which I understand still exists today.

Viewed from the other end of a snowbound Ropley station, ex-SR S15 Class 4-6-0 No. 30837 and U Class 2-6-0 No. 31639 pass with the LCGB S15 Commemorative Tour on 16 January 1966. This tour marked the last run of the S15 Class and was so heavily subscribed that a repeat tour for late-booking passengers had been run a week earlier.

It was comparatively rare to see an eastbound diversion double-headed, but on 22 May 1966 ex-BR Class 4MT 2-6-4T No. 80139 assists ex-SR Class MN 4-6-2 No. 35029 *Ellerman Lines* through Ropley with the 09.54 Weymouth–Waterloo. No. 35029 is 'preserved' as a sectioned exhibit in York Railway Museum.

Ex-LMS Class 5 4-6-0 No. 45493 passes through Arlesford station with the 08.55 Bournemouth–Waterloo on 22 May. This engine would have worked into Bournemouth on the Saturday evening with the through train from York. Since the balancing working was not until Monday, the engine was available for Bournemouth shed to use on a Sunday diagram.

Southern By-Ways: Branch Lines of BR's Southern Region in the 1960s

A week earlier, on 15 May 1966, and in far better weather, ex-LMS Class 5 4-6-0 No. 45493 passes through Itchen Abbas with the 08.55 Bournemouth–Waterloo. Interestingly, on both occasions it was this same, quite clean, locomotive.

Ex-SR WC Class 4-6-2 No. 34019 *Bideford* passes through Itchen Abbas with the 16.30 Waterloo–Weymouth on 1 May 1966. As at Ropley, by 1965 the passing loop had been removed, leaving Arlesford and Medstead as the only passing places.

Salisbury–Fareham–Southampton

The line from Andover, via Stockbridge, joined the Salisbury to Romsey line here at Kimbridge Junction. The line closed in 1965, but for a time was used to store this withdrawn electric unit. Unit No. 4348 was one of fifty former Eastern Section three-car suburban units augmented with a fourth steel curved-bodied coach. Vandals have clearly been at work.

Ex-SR U Class 2-6-0 No. 31639 and Q1 Class 0-6-0 No. 33006 is approaching Kimbridge Junction with an earlier part of the LCGB Wilts and Hants tour of 3 April 1966. No. 31639 was built as new, rather than being rebuilt from River Class tanks.

Ex-SR S15 Class 4-6-0 No. 30824 approaches Romsey with a lengthy freight from Salisbury. This Salisbury-based loco was built in 1927 by Maunsell as a modification of an original design by Urie in 1920.

An unusual sight at Botley of the *Bournemouth Belle*, diverted via Fareham as a result of engineering works in preparation for the electrification of the Bournemouth main line. The engine is an ex-SR rebuilt BB Class 4-6-2 still carrying its squadron nameplate and crest.

Ex-LNER A3 Class 4-6-2 No. 60103 *Flying Scotsman* passes through Botley with the return leg of a Loco Preservation (Sussex) rail tour on 17 September 1966. No. 60103, already preserved in LNER livery, had come onto Southern territory to work a couple of rail tours.

Ex-SR Q1 Class 0-6-0 No. 33027 hauls a freight from Eastleigh towards Fareham at Knowle Junction. This section of track was the end of the now closed Meon Valley branch, the start of which was illustrated on p. 48, and was currently in use for southbound traffic only.

Shortly beyond where the previous image was taken, the line passed through Knowle Tunnel. Ex-SR WC Class 4-6-2 No. 34093 *Saunton* bursts out of the single-line tunnel on the approach to Fareham with the diverted 13.30 Waterloo to Bournemouth and Weymouth in November 1964.

Ex-SR WC Class 4-6-2 No. 34002 *Salisbury*, still carrying nameplate and crest, is arriving at Fareham with the all-Pullman 12.30 Waterloo–Bournemouth *Bournemouth Belle* on 8 November 1964, diverted via Fareham due to electrification works between Eastleigh and Southampton.

A double-track deviation for Knowle Tunnel was built in 1906, but a large landslip in 1962 caused the northbound line to be taken out of use. The Down southbound line was slewed at both ends to make a single deviation line for the exclusive use of northbound traffic, with southbound traffic exclusively using the tunnel line. Here, ex-SR WC Class 4-6-2 No. 34047 *Crediton* is hauling a northbound Bournemouth to Waterloo train on this deviation.

Ex-SR BB Class 4-6-2 No. 34079 *141 Squadron* is passing through Swanwick with the diverted 13.29 Bournemouth–Waterloo on 8 November 1964.

BR Class 4MT 2-6-0 No. 76006 is passing through Swanwick with the 11.30 Bristol to Portsmouth on 8 November 1964.

Viewed through the arches of the A3051 bridge, ex-SR WC Class 4-6-2 No. 34104 *Bere Alston* passes through Swanwick station with the diverted 09.30 Waterloo–Bournemouth on 8 November 1964.

Hayling Island Branch

Ex-LBSCR A1X Class 0-6-0 No. 32670 passes the Hayling Island home signal with a train from Havant on 8 September 1963. This was one of a class of fifty locos built in the 1870s to a Stroudley design, designated 'A1'.

After running round its train, ex-LBSCR A1X Class 0-6-0 No. 32670 is waiting in the bay platform at Hayling Island with some enthusiasts chatting to the engine crew.

On 8 September 1963, ex-LBSCR A1X Class 0-6-0 No. 32670 returns its train across the flats near North Hayling, heading for Havant. No. 32670 was one of twenty of the A1 Class rebuilt by Marsh from 1911 with new boilers and extended smokeboxes thus designated A1X.

Ex-LBSCR A1X Class 0-6-0 No. 32670 is crossing Langstone Viaduct bound for Hayling Island on 26 October 1963. This timber trestle viaduct was the major engineering feature on the branch, and because of weight restrictions the A1X Class was the only type allowed on the branch.

Ex-LBSCR A1X Class 0-6-0 No. 32650 runs the short distance between Langstone and the viaduct on the 14.20 from Havant on 2 November 1963 – the penultimate day of services on the branch.

In front of many interested spectators, ex-LBSCR A1X Class 0-6-0 No. 32650 is running round its train at Hayling Island station on 2 November 1963.

Ex-SR U Class 2-6-0 No. 31791 is arriving at Havant from Fratton with the LCGB Hayling Farewell tour on Sunday 3 November 1963. This loco was rebuilt from River Class 2-6-4T A791 *River Adur*.

Ex-LBSCR A1X Class 0-6-0 No. 32636 is heading for Hayling Island near Langstone with the Hayling Farewell tour on 3 November 1963. At the time, this was the oldest working steam locomotive still in BR service.

The sun sets over Langstone Harbour, and the Hayling Island branch, as ex-LBSCR A1X Class 0-6-0T No. 32650 crosses the timber viaduct linking Hayling Island with the mainland with the 16.20 Hayling Island–Havant on Saturday 2 November 1963 – the last day of regular passenger service.

The final sunset! Ex-LBSCR A1X Class 0-6-0 No. 32662 crosses Langstone Viaduct with a service from Havant to Hayling Island on Saturday 2 November 1963. It was due to the state of this viaduct and the cost of replacement that the branch was closed.

Lymington Branch

As was normal practice, ex-LSWR M7 Class 0-4-4T No. 30480 arrives on the outer face of the Up platform at Brockenhurst with a train from Lymington – in this case, the 18.29 arrival. Most branch trains were timed to connect with Up main-line services – here the 18.20 from Bournemouth, scheduled to arrive 11 minutes later.

Having discharged passengers, at a suitable space in traffic the auto-train from Lymington would reverse out of the Up platform and cross over into the outer face of the Down platform ready to receive passengers travelling to Lymington. On a different occasion, ex-LSWR M7 0-4-4T No. 30107 is completing the same manoeuvre.

Lymington trains ran about three-quarters of a mile in a south-westerly direction from Brockenhurst station on the main line to Bournemouth to reach Lymington Junction, where they diverged southwards on the branch line to Lymington. Ex-LSWR M7 Class 0-4-4T No. 30107 has just come off the branch and is heading on the Up main line to Brockenhurst in March 1964. Lymington Junction signal box can be seen in the background.

On the branch, the 16.18 from Lymington has just crossed under a farm track and footpath at Battramsley in August 1963, hauled by ex-LSWR M7 Class 0-4-4T No. 30480. The air reservoir below the overhang and Westinghouse pump on the side of the smokebox are clearly visible.

In normal service, one push-pull train was sufficient to work trains between Lymington and Brockenhurst. Later on the same day as the previous image, we again see ex-LSWR M7 Class 0-4-4T No. 30480 approaching Battramsley with the 17.25 from Lymington.

On another day in August 1963, ex-LSWR M7 Class 0-4-4T No. 30052 is providing the 18.16 service from Lymington, heading across Setley Plain. The coaching stock comprises old Maunsell stock converted to push-pull working with a driving compartment at the end.

Lymington boasts two stations about half a mile apart – Town and Pier Head. In this view, ex-LSWR M7 Class 0-4-4T No. 30052 is just leaving the Town station for Brockenhurst, with the train pictured in the next image.

This view of Pier Head station shows how conveniently placed it is to facilitate transfer to the Isle of Wight ferry to Yarmouth. With the ferry in the right of the picture, ex-LSWR M7 Class 0-4-4T No. 30052 waits the departure time of the 17.25 for Brockehurst.

With the withdrawal from service of the last M7 tanks, the Lymington service was taken over by ex-LMS Ivatt 2-6-2Ts and BR 2-6-4Ts. In April 1967, No. 41312 is seen departing Brockenhurst. After the withdrawal of the M7s, the Maunsell-era push-pull-fitted stock was gradually phased out.

Ex-LMS Ivatt 2-6-2T No. 41312 crosses the road that ran across the southern edge of the New Forest towards Sway and Burley in April 1967 – the boundary of the New Forest is about a mile to the south.

The line between Lymington Pier Head and Town stations initially runs alongside the Lymington River, and ex-LMS Ivatt 2-6-2T No. 41320 is viewed from the marina on the opposite side with a train for Brockenhurst on the last day of steam working on Sunday 2 April 1967.

The line then crosses the Lymington River to enter the Town station. Ex-LMS Ivatt 2-6-2T No. 41320 is crossing the river bridge on the last day of steam working, 2 April 1967. The headboard reads 'The Last Steam Branch 1967'. From this date, the branch has been worked by modern traction.

Somerset and Dorset Line

Ex-LMS 4F 0-6-0 No. 44560 is passing Bailey Gate with a southbound freight in April 1964. This was one of five 4Fs (Nos 44557–44561) built specially for the SDJR.

BR Class 5 4-6-0 No. 73088 arrives at Blandford Forum with the 16.13 from Evercreech Junction. Note the tall signal box behind the engine, which was built to replace the original on the other platform after it had been destroyed by a lightning strike.

Ex-BR 5MT 4-6-0 No. 73052 departs Shillingstone with the 15.35 mixed train from Templecombe. No. 73052 was a long-time resident of Bath Green Park shed, being one of three of this class built with larger tenders for working over the SDJR.

Ex-GWR Collet-designed 0-6-0 No. 2217 works a southbound freight near Templecombe, a sign of the takeover of the northern part of the SDJR by the Western Region of British Railways in 1958.

BR 5MT 4-6-0 No. 73001 leaves Cole with the 09.45 from Highbridge to Templecombe in January 1966. Just before Cole station, it would have passed over the Great Western main line from Frome to Taunton.

An unidentified BR 4MT 2-6-4T crosses Charlton Viaduct at Shepton Mallet with the 07.00 Templecombe to Bath Green Park local passenger train on the last morning of regular service, 5 March 1966.

Another unidentified BR 4MT 2-6-4T is approaching the summit of the 1 in 50 climb to Masbury Summit on 3 March 1966. Although more than 10 miles away, the unmistakable shape of Glastonbury Tor is visible in the distance.

With just another half mile to go to reach the summit, BR 4MT 2-6-4T No. 80034 is passing Masbury Halt with the 09.05 from Templecombe to Bath Green Park on 5 March 1966.

Heading south, ex-LMS 8F Class 2-8-0 No. 48767 is leaving Binegar with the 09.53 from Bath Green Park to Bournemouth on 5 March 1966. It has another mile or so of 1 in 63 to reach Masbury Summit.

Ex-SDJR 7F Class 2-8-0 No. 53806 is slogging up the 1 in 50 grade through Midsomer Norton South in July 1964 with the 14.00 freight from Bath Green Park to Evercreech Junction and Templecombe. This station is preserved by the Somerset & Dorset Railway Heritage Trust.

An unidentified 'Jinty', ex-LMS 3F 0-6-0T, is providing banking assistance for No. 53806 as it hauls its freight up the bank to Masbury. Radstock shed had an allocation of these tanks for banking and other duties.

BR 4MT 2-6-4T No. 80043 has just emerged from the twin-bore tunnel north of Chilcompton with the 16.35 from Bath Geen Park to Templecombe on 3 March 1966 on the climb towards Masbury Summit.

A lone ex-LMS 'Jinty' 3F 0-6-0T, probably No. 47276, heads north from Radstock shortly after a passing storm. It is heading to work the colliery at Writhlington on the morning of 4 March 1966.

Ex-LMS 'Jinty' 3F 0-6-0T No. 47276 is taking a freight, mainly consisting of coal empties, towards Writhlington on the afternoon of 4 March 1966.

Later in the morning on 4 March 1966, ex-LMS 'Jinty' 3F 0-6-0T No. 47276 is shunting the upper sidings at Writhlington Colliery.

As can be judged from the relative height of the signal box on the main line, there was a sharp climb from the pithead to the transfer sidings. Ex-LMS 'Jinty' 3F 0-6-0T No. 47276 is charging the incline with one of a number of small loads to the transfer sidings at the top.

A BR 4MT 2-6-4T and ex-LMS Ivatt 2-6-2T double-head the 12.00 Templecombe to Bath Green Park on 4 March 1966. As it is the penultimate day of services, the double-heading is obviously a means of getting an additional engine from Templecombe or Radstock to Bath Green Park shed.

An unidentified BR 4MT is climbing over Tucking Mill Viaduct towards Combe Down Tunnel on 4 March 1966 with the 15.05 local passenger train from Templecombe to Bath Green Park.

Ex-SDJR 7F 2-8-0 No. 53807 and ex-LMS/SDJR 4F 0-6-0 No. 44558 work a Home Counties Railway Society tour across Prestleigh Viaduct on the 1 in 50 approach to Shepton Mallet on 7 June 1964.

Ex-SDJR 7F 2-8-0 No. 53807 and ex-LMS/SDJR 4F 0-6-0 No. 44558 being serviced on the shed at Bath Green Park after working the HCRS rail tour.

The HCRS rail tour was worked on from Bath Green Park by ex-BR Castle Class 4-6-0 No. 7023 *Penrice Castle*, photographed between Bath and Mangotsfield. A Castle Class loco on this stretch of line would have been very uncommon in normal service.

Two special trains were run in opposite directions on the S&D on the last Saturday of operation, 5 March 1966. Ex-LMS 8F 2-8-0 No. 48706 is near Binegar on the southbound climb to Masbury Summit with the Great Western Society 'last day special' from Bath Geen Park to Bournemouth.

Ex-LMS 8F 2-8-0 No. 48706 passes Pitcombe church, near Cole, on the continuation southwards of the Great Western Society special to Bournemouth on 5 March 1966.

In the opposite direction, on 5 March 1966, the LCGB 'last day special' was worked northwards from Evercreech Junction by ex-SR WC/BB Class 4-6-2 Nos 34006 *Bude* and 34057 *Biggin Hill*, seen here at Prestleigh, south of Shepton Mallet. The elongated smoke deflectors – unique to *Bude* – can be seen to good effect.

Ex-SR WC/BB Class 4-6-2 Nos 34006 *Bude* and 34057 *Biggin Hill* are coasting downhill on the approach to Chilcompton with the LCGB 'last day special'. The wartime tank traps are still in place.

The very last day, Sunday 6 March 1966, again saw two special trains running in opposite directions. Running south from Bath Green Park in much poorer weather than on the previous day is the Stephenson Locomotive Society special from Bath Green Park hauled by ex-LMS Class 8F No. 48706 and BR 4MT 2-6-4T No. 80043, seen near Binegar on the climb to Masbury Summit.

The northbound special, organised by the RCTS, is passing Templecombe Lower with ex-SR MN Class 4-6-2 No. 35028 *Clan Line*, viewed from the Southern main line. Merchant Navy Class locos passing over the S&D line at Templecombe were ten a penny, but one passing under the main line was extremely rare. The lower platform at Templecombe was used only for the occasional late evening train; most trains reversed at Templecombe Junction just beyond up into the main-line platform.

The end of the Somerset & Dorset! Work is taking place to slew the southbound track to make a junction with the line through Radstock West to Bristol to allow trains from Writhlington Colliery to transport coal to Portishead power station. Writhlington Colliery finally closed in September 1973.

Although the bulk of the SDJR closed on 6 March 1966, a spur from Broadstone Junction to Blandford Forum remained open until 1969. On 16 October 1966 an LCGB rail tour to Hampshire is seen crossing a rather swollen River Stour hauled by BR Class 3MT 2-6-0 No. 77014 on its return from Blandford Forum.

BR Class 3MT 2-6-0 No. 77014 is seen leaving Spetisbury after having a photographic stop there on the LCGB Hampshire rail tour with BR Class 4MT 2-6-0 No. 76026 on the back. Because of its unique appearance in the south of the country, No. 77014 was a popular choice for rail tours at this time.

Swanage Branch

Because a direct line from Wareham would have breached the town's ancient walls, the junction for Swanage was sited at the hamlet of Worgret, about a mile south-west of Wareham. At the junction, the branch turned at right angles to head south-eastwards. Ex-LSWR M7 Class 0-4-4T No. 30053 is propelling its train towards the junction with the route cleared to take the branch.

Ex-LSWR M7 Class 0-4-4T No. 30053 passes Furzebrook with the 13.40 Swanage–Wareham in April 1964. The head-shunt for the ball-clay siding is on the right. This had connected with a narrow-gauge railway, which had closed in 1957, serving clay workings in the Purbeck Hills.

Ex-LSWR M7 Class 0-4-4T No. 30056 propels its train away from Corfe Castle station on its way to Swanage in the summer of 1963 with the 16.57 from Wareham.

Ex-LSWR M7 Class 0-4-4T No. 30053 leaves Swanage with the 13.40 auto-train to Wareham in April 1964. This scene could still be created today (without the auto-coaches) since this loco, restored from its forlorn condition portrayed here, currently operates on the preserved Swanage Railway.

No. 30053 at rest in front of the single-line shed at Swanage before working the train illustrated above. As with the Lymington branch, all remaining M7s were withdrawn from service at Bournemouth shed in early May 1964.

Viewed through the arches of the main A352 road from Wareham to Dorchester, the 'new order' in the form an unidentified BR Class 4MT 2-6-4T approaches Worgret Junction with a Swanage branch train.

Southern By-Ways: Branch Lines of BR's Southern Region in the 1960s

At Worgret Junction, the signalman hands the single-line token to the crew of ex-LMS Ivatt 2-6-2T No. 41230 in the summer of 1966.

An early morning train from Swanage approaches Worgret Junction across the plain formed by the River Frome hauled by an unidentified ex-LMS Ivatt 2-6-2T. The Purbeck Hills are in the distance.

An ex-LMS Ivatt 2-6-2T crosses the River Frome as it makes its way on the approach to Worgret Junction on the morning of Saturday 11 July 1966, the day of England's World Cup football win.

Corfe Castle, built by William the Conqueror in the eleventh century, commands a gap in the Purbeck Hills between Wareham and Swanage. Not only is it a prominent feature in various photographs of the branch, but it also provides a good vantage point to see an ex-LMS Ivatt 2-6-2T leaving the village station bound for Wareham.

Southern By-Ways: Branch Lines of BR's Southern Region in the 1960s 91

From a lower angle on the castle hill, an ex-BR Class 4MT 2-6-4T leaves the station with the 16.20 from Swanage in April 1966.

On summer Saturdays the occasional through train worked from Swanage to Waterloo. With the village of Corfe Castle spread out behind, a Bullied Light Pacific pauses for passengers at the station with the 13.23 on 11 July 1966.

The cloud almost gives a ghostly aura to Corfe Castle in the background as ex-LMS 2-6-2T No. 41316 stands at the station with the 14.00 from Wareham on 11 July 1966.

Ex-LMS 2MT 2-6-2T Nos 41284 and 41301 pass Corfe Castle on the LCGB Dorset Belle tour on 27 February 1966 – viewed from West Hill after an exhausting run up the slope during the 10-minute photographic stop at Corfe Castle station.

Southern By-Ways: Branch Lines of BR's Southern Region in the 1960s

Prior to the photographic stop at Corfe Castle, ex-LMS 2MT 2-6-2T Nos 41284 and 41301 head towards Harman's Cross on the LCGB Dorset Belle tour on 27 February 1966.

A rebuilt Bullied Light Pacific approaches Harman's Cross with the 11.20 through train from Swanage to Waterloo.

Ex-SR rebuilt WC Class 4-6-2 No. 34004 *Yeovil* is near Furzebrook on a Warwickshire Railway Society special train to Swanage on 11 June 1967, just a month before the elimination of steam on the Southern Region.

Ex-SR rebuilt WC Class 4-6-2 No. 34004 *Yeovil* continues its journey to Swange with the Warwickshire Railway Society special train on 11 June 1967. With the ruins of Corfe Castle prominent in the background, this classic view is still obtainable with steam trains on the heritage Swanage Railway.

To conclude: then and now. In April 1964, ex-LSWR M7 Class 0-4-4T No. 30053 is hauling the branch train from Swanage towards Harman's Cross.

In August 2013 the same M7, No. 30053, in a somewhat more presentable external condition, is hauling a train from Swanage to Norden (a half-mile or so beyond Corfe Castle) just beyond Harman's Cross. Today's heritage Swanage Railway runs trains between Swanage and Norden, with No. 30053 as one of their flagship locos.